The regulation and use of radioisotopes in today's world

Introduction

More than 100 years ago, scientists discovered that many elements commonly found on earth occur in different atomic configurations. These varying configurations, called **isotopes**, were found to have identical electronically charged particles and identical chemical properties, but different atomic weights and physical properties.

It was soon discovered that some isotopes of elements were radioactive. The dense central portion (called the **nucleus**) of an atom of the element emits energy in several different forms. Radioisotopes are simply atoms with nuclei that are seeking a more stable nuclear configuration by emitting radiation. Scientists have learned that more radioisotopes could be created by subjecting certain elements to radiation inside a nuclear reactor or bombarding them using a particle accelerator.

Gradually we have learned to harness these radioisotopes for use in our modern, high-tech world. In this brochure are described some of the most common uses for radioisotopes, as well as the relative benefits and hazards involved in their applications. The appendix at the end of this brochure describes various uses of radioisotopes in this country.

The regulation and use of radioisotopes in today's world

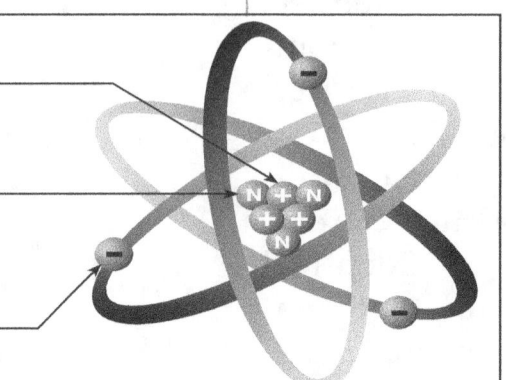

INSIDE THE ATOM

Protons ...are the nucleus which consists of protons, which carry positive electric charges...

Neutrons ...and neutrons, which carry no electric charge.

Electrons Encircling the nucleus are electrons which carry negative electric charges.

Also explained are the properties of radiation that enable it to change the physical characteristics of certain materials. Because they emit tell-tale ionizing radiation, extremely small quantities of radioisotopes can be traced and measured using special equipment. This property makes them a useful diagnostic tool in medicine. The radiation from some radioisotopes can penetrate thick metal parts and provide a way to "see" inside objects that are impenetrable to light. Radiation deposits sufficient energy in human tissue to disrupt normal cell function as it passes through, thus providing a unique method of attacking and destroying cancerous cells and tumors. Radiation can also be used to kill bacteria and germs that contaminate medical instruments and some foodstuffs.

Finally, this brochure describes the responsibilities of the U.S. Nuclear Regulatory Commission (NRC), some other Federal agencies, and the States, in regulating the manufacture, use, and possession of radioactive materials.

The regulation and use of radioisotopes in today's world

A few radioisotopes occur naturally but most are man made. A radioisotope is typically described by its name followed by a number, such as carbon-14 (C-14) or fluorine-18 (F-18). The number represents the atomic weight or the total number of protons and neutrons that make up the atom's nucleus.

Radioisotopes have unique properties that make them useful tools in solving problems.

Three predominant types of radiation are emitted by radioisotopes: (1) alpha particles, (2) beta particles, and (3) gamma rays. The different types of radiation can penetrate materials of varying thicknesses such as paper, body tissue, or concrete. A single sheet of paper will stop an **alpha particle**, which, because it is heavy, has very little penetrating power. Alpha emitters are typically used in smoke detectors. A thin sheet of metal will stop a **beta particle**, which,

Why Are Radioisotopes Useful?

1. *Radioisotopes emit different kinds of radiation.*

Penetrating power of radiation

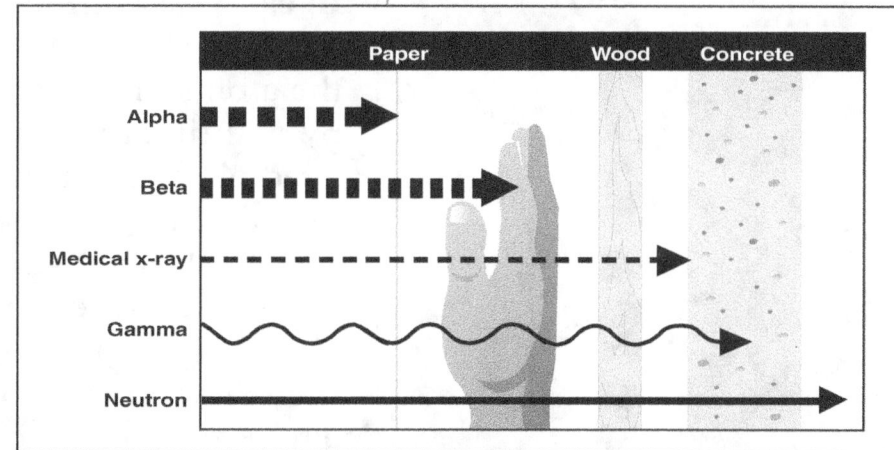

The regulation and use of radioisotopes in today's world

because it is lighter, is more penetrating than an alpha particle. Beta emitters such as strontium-90 (Sr-90) are used in the treatment of eye disease. **Gamma rays**, which have no charge or weight, can be extremely energetic and highly penetrating. Several **meters** of concrete or several millimeters of other dense materials, such as lead, are needed to block gamma rays. Radioisotopes that emit gamma rays are used frequently in medical applications against tumors, and in industry to check for cracks or flaws in valves and piping. Because of their varying penetrating properties, radioisotopes can be manipulated to perform different tasks.

2. *The length of a radioisotope's life is predictable.*

The process of radioactive decay, in which radioisotopes lose their radioactivity over time, is measured in **half-lives**. A half-life of a radioactive material is the time it takes one-half of the atoms of the radioisotope to decay by emitting radiation. The half-life of a radioisotope can range from fractions of a second (radon-220) to millions of years (thorium-232). When radioisotopes are used in industry or medicine, it is vital to know how rapidly they disappear, so that the precise amount of radioisotope available for the medical procedure or industrial use is known.

The regulation and use of radioisotopes in today's world

We can trace the movement of a chemical element by using a radioisotope of that element. For instance, the chemical element iodine concentrates naturally in the thyroid. By using a radioactive isotope of iodine as a tracer, a picture can be taken of the material's path through the body and its deposit in a specific organ.

3. *Radioisotopes of a chemical element behave in the same manner as a stable, nonradioactive element.*

Half-life decay

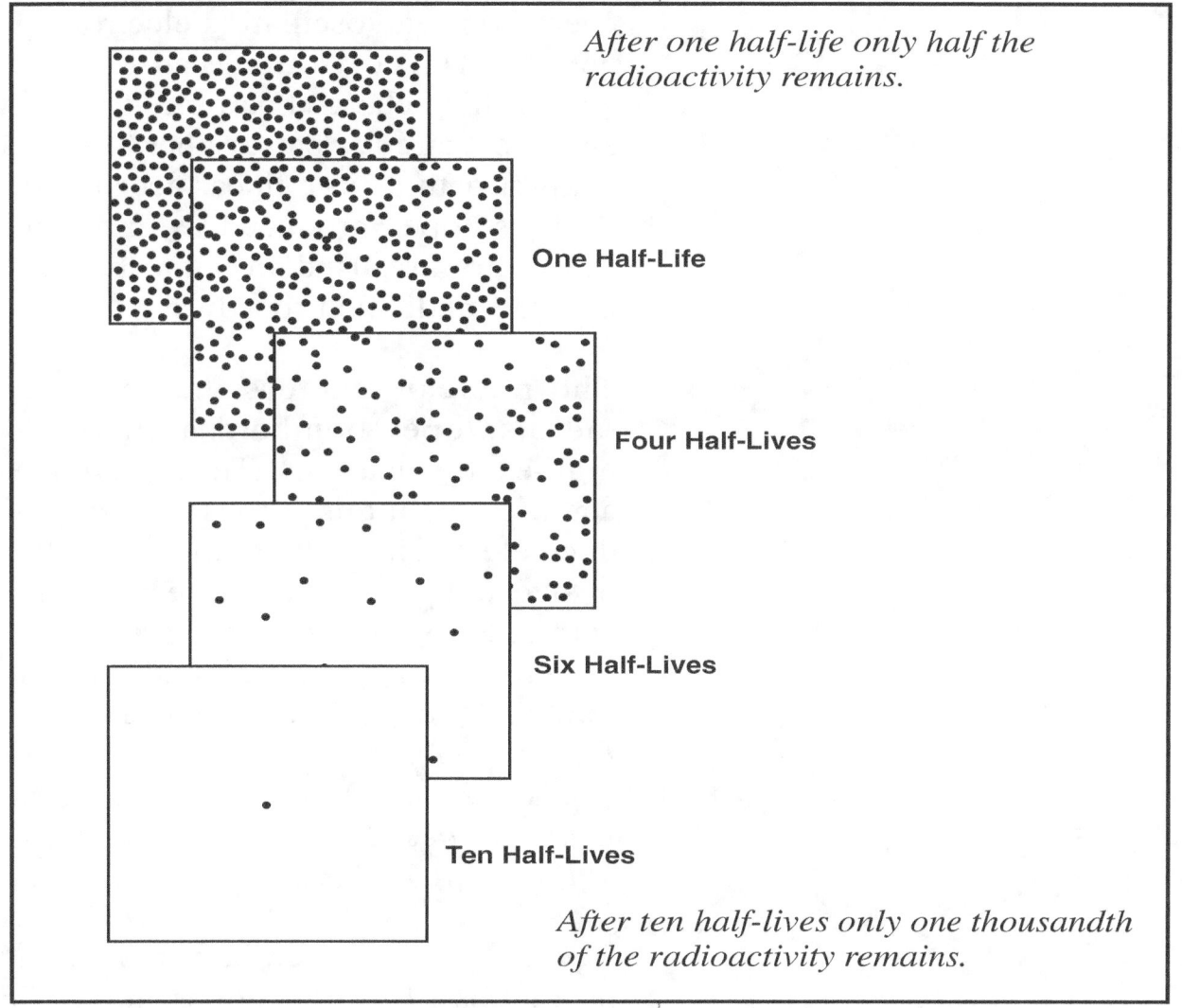

After one half-life only half the radioactivity remains.

One Half-Life

Four Half-Lives

Six Half-Lives

Ten Half-Lives

After ten half-lives only one thousandth of the radioactivity remains.

The regulation and use of radioisotopes in today's world

Where Do Radioisotopes Come From and Who Regulates Their Use?

Radioisotopes come from three sources: (1) from nature, such as radon in the air or radium in the soil; (2) from machine-produced nuclear interactions in devices, such as **linear accelerators and cyclotrons;** or (3) from nuclear reactors.

A linear accelerator is a long, straight tube in which charged particles gain energy through oscillating electromagnetic fields. A cyclotron is an accelerator in which charged particles travel in an almost circular path, rather than in a straight path as in a linear accelerator. Radioisotopes produced in linear accelerators are used in some modern nuclear medicine procedures.

The nuclear reactors that produce radioisotopes bombard atoms with high-energy neutrons. The research reactors used for this purpose do not produce electricity and are much smaller in size and power than large power reactors. Reasearch reactors are mostly used for training and for identifying the composition of certain elements. *Forty-seven research reactors are licensed by the NRC to operate in the United States. They are located mostly at large universities.*

Because of the potentially hazardous properties of radioisotopes, their use must be closely regulated to ensure that public health and safety are protected.

The regulation and use of radioisotopes in today's world

The licensing and regulation of radioisotopes in the United States are shared by the NRC, the U.S. Environmental Protection Agency (EPA), and many State governments. The EPA is also responsible for, among other things, setting air emission and drinking water standards for radionuclides. The States regulate radioactive substances that occur naturally or are produced by machines, such as linear accelerators or cyclotrons. The Food and Drug Administration (FDA) regulates the manufacture and use of linear accelerators; the States regulate their operation.

The NRC is the Federal agency given the task of protecting public health and safety and the environment with regard to the safe use of nuclear materials. Among its many responsibilities, the NRC regulates medical, academic, and industrial uses of nuclear materials generated by or from a nuclear reactor. Through a comprehensive inspection and enforcement program, the NRC ensures that these facilities operate in compliance with strict safety standards.

The NRC has relinquished its authority to regulate certain radioactive materials, including some radioisotopes, to most of the States. These States, which have entered into an agreement assuming this regulatory authority from

What Is the Role of the Nuclear Regulatory Commission?

The regulation and use of radioisotopes in today's world

the NRC, are called **Agreement States**, and are shown on the map below. Agreement States, like the NRC, regulate reactor-produced radioisotopes within their borders and must provide at least as much health and safety protection as the NRC.

The NRC maintains approximately 6,000 licenses for the use of radioactive materials, and the Agreement States maintain approximately 16,000

Map of Agreement States

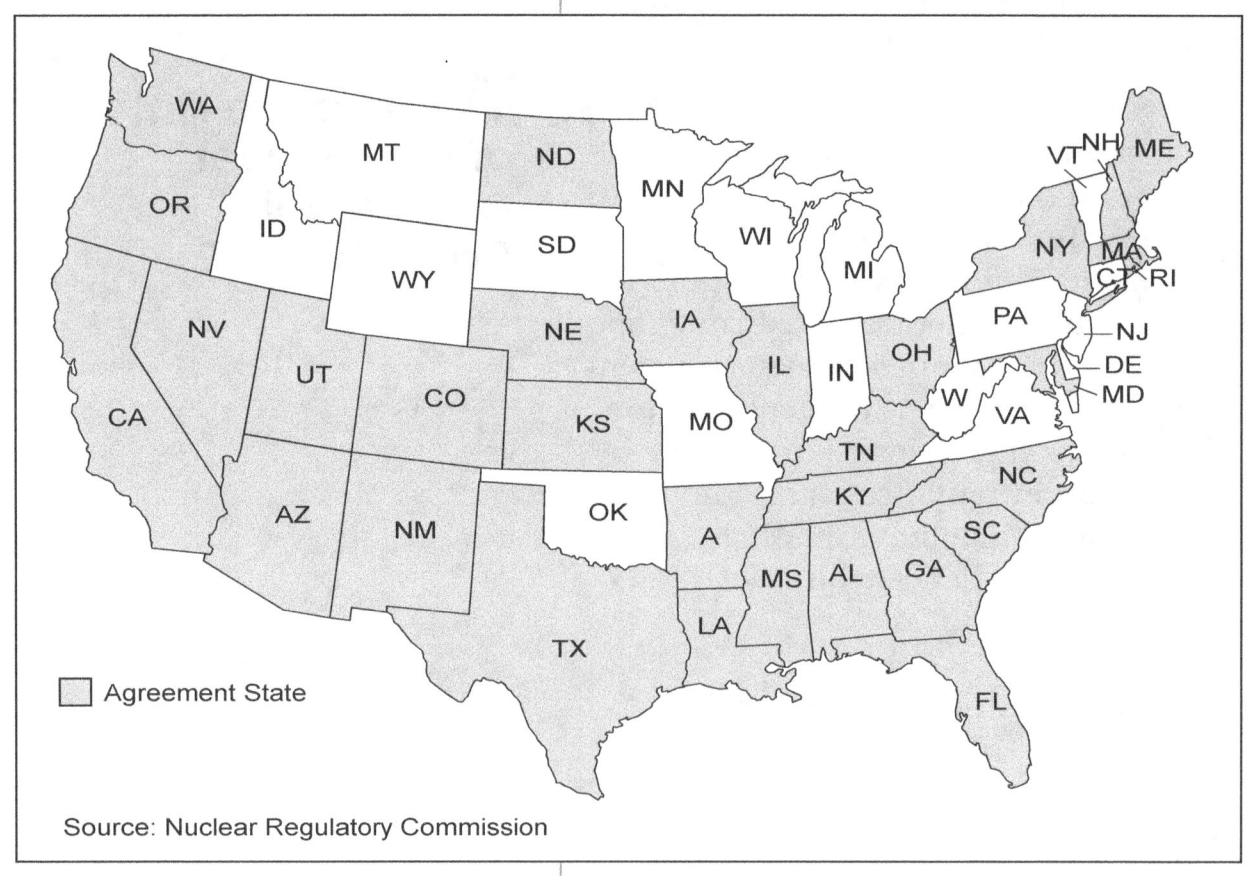

Source: Nuclear Regulatory Commission

The regulation and use of radioisotopes in today's world

materials licenses. Every license specifies the type, quantity, and location of radioactive material that may be possessed and used. When radioactive material is transported, special packaging and labeling are required. Also specified in each license are the training and qualification of workers using the materials, specific procedures for using the materials, and any special safety precautions required. Every licensee is inspected periodically either by the NRC or the Agreement State to ensure that radioactive materials are being used and transported safely. Violators of regulatory requirements are subject to fines and other enforcement actions, including loss of license.

The regulation and use of radioisotopes in today's world

How Are Radioisotopes Used in Medicine?

About one-third of all patients admitted to U.S. hospitals are diagnosed or treated using radioisotopes. Most major hospitals have specific departments dedicated to radiation medicine. In all, about 112 million nuclear medicine or radiation therapy procedures are performed annually, with the vast majority used in diagnoses. Radioactive materials used as a diagnostic tool can identify the status of a disease and minimize the need for surgery, reducing the risks from postoperative infection.

Diagnostic Applications

Radioisotopes give doctors the ability to "look" inside the body and observe soft tissues and organs, in a manner similar to the way x-rays provide images of bones. Radioisotopes carried in the blood also allow doctors to detect clogged arteries or check the functioning of the circulatory system. Some chemical compounds concentrate naturally in specific organs or tissues in the body. For example, iodine collects in the thyroid while various compounds of technetium-99m* (Tc-99m) collect in the bones, heart, and other organs. Taking advantage of this proclivity, doctors can use radioisotopes of these elements as **tracers.** A radioactive tracer is chemically attached to a compound that will

---

*The "m" refers to "metastate," which pertains to the isotope's short half-life and elevated energy state.

concentrate naturally in an organ or tissue so that a picture can be taken. The process of attaching a radioisotope to a chemical compound is called **labeling**.

To detect problems within a body organ, doctors use **radiopharmaceuticals** or radioactive drugs. Radioisotopes that have short half-lives are preferred for use in these drugs to minimize the radiation dose to the patient. In most cases, these short-lived radioisotopes decay to stable elements within minutes, hours, or days, allowing patients to be released from the hospital in a relatively short time.

The radioisotope used in about 80 percent of nuclear diagnostic procedures is Tc-99m. The penetrating properties of its gamma rays and its short (6-hour) half-life help reduce risk to the patient from more prolonged radiation exposure.

Because of their short half-lives, certain radiopharmaceuticals must be produced, shipped to the hospital, and then used within a couple of weeks. Short-lived radionuclides such as Tc-99m, gallium-67, and thallium-201 are often used to diagnose the functioning of

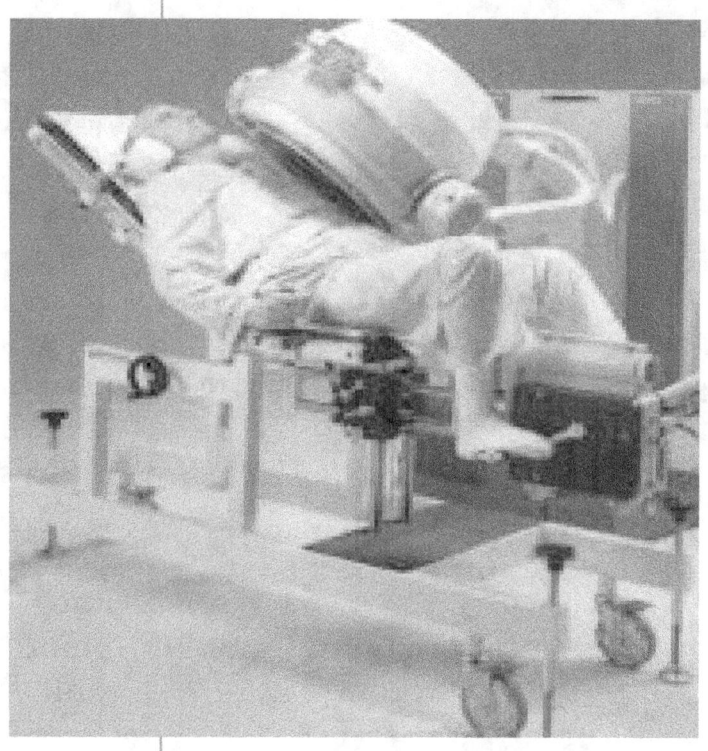

Gamma camera used to detect dysfunctional activity in body organs

The regulation and use of radioisotopes in today's world

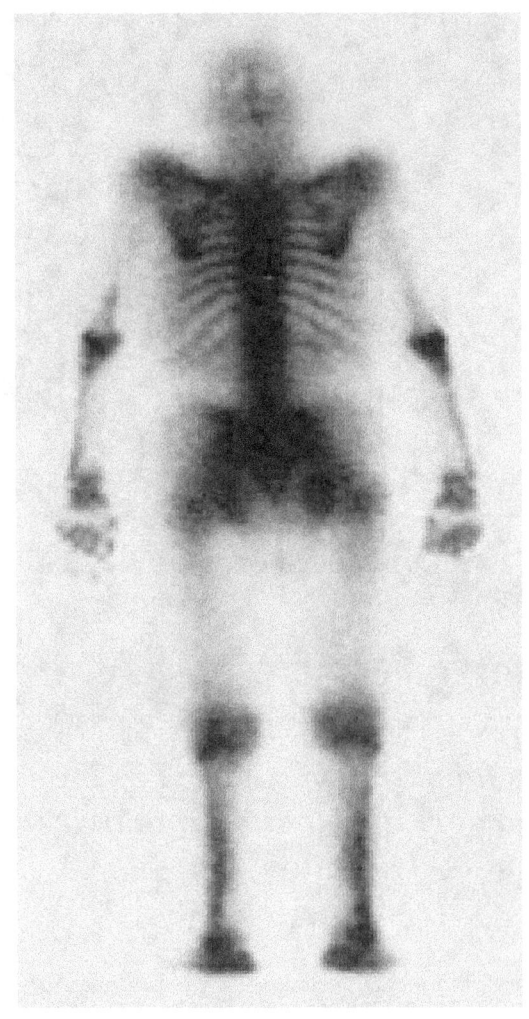

Tracer used in the body of a 65-year old man indicating bone density problems as a result of a thyroid condition

Photo: Courtesy of Appleton & Lange, from "A Clinical Manual of Nuclear Medicine," by John Walker and Donald Margouless, 1984.

the heart, brain, lung, kidney, or liver. For example, Tc-99m is used to diagnose osteoporosis, a condition caused by calcium deficiency in older people, especially women.

To evaluate the presence of heart disease, a radioisotope is injected into a patient's bloodstream while he or she is exercising on a treadmill. The radioisotope travels toward the heart, allowing doctors to follow the blood flow on a screen. While looking at the image, doctors can check for reduced blood flow through the arteries, a possible signal of heart disease.

Nuclear imaging is also used to evaluate brain function. Organic radiochemicals are labeled with F-18 and then injected into the bloodstream. A device called a **gamma camera** detects radiation emitted from the organ, displaying an image that can enable the physician to detect blockages or other dysfunctional activity.

For some diagnostic tests, the patient need not come into contact with radioactivity at all. The tests are performed on blood or other fluids taken from the patient, using a procedure called **radioimmunoassay**. These tests can detect some diseases by identifying and measuring the amounts of hormones, vitamins, enzymes, or drugs in the body.

The regulation and use of radioisotopes in today's world

Therapeutic Applications

The same property that makes radiation hazardous can also make it useful in helping the body heal. When living tissue is exposed to high levels of radiation, cells can be destroyed or damaged so they can neither reproduce nor continue their normal functions. For this reason radioisotopes are used in the treatment of cancer (which amounts to uncontrolled cell division). Although some healthy tissue surrounding a tumor may be damaged during the treatment, mostly cancerous tissue can be targeted for destruction.

A device called a **teletherapy unit** destroys malignant tumors with gamma radiation from a radioisotope such as cobalt-60 (Co-60). Teletherapy units use a high-energy beam of gamma rays to reduce or eradicate tumors deep within the body. These units are licensed by the NRC because they use byproduct material that is produced only by a nuclear reactor.

Another treatment, called **brachytherapy**, destroys cells by placing the radioisotope (in the form of a sealed source) directly into the tumor. Generally, two techniques are used for this type of treatment: (1) direct, manual implantation of a radiation source by a physician or (2) automated implantation using a device called a **remote afterloader**. The NRC as well as

The regulation and use of radioisotopes in today's world

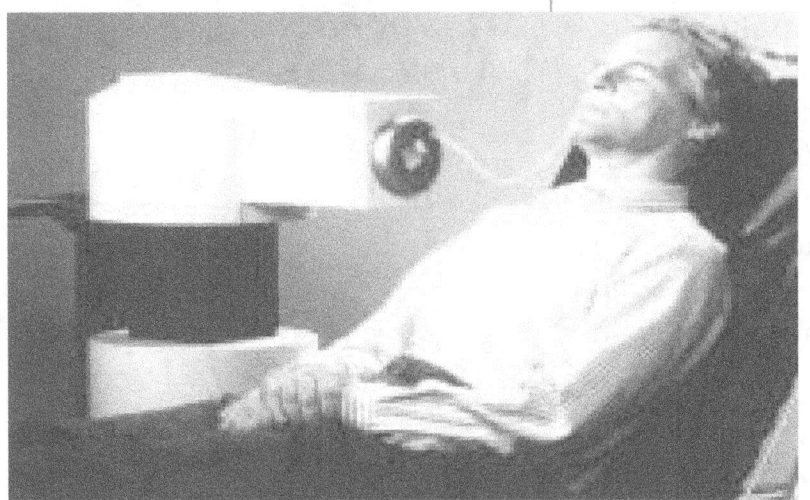

Brachytherapy device

Photo: Courtesy of Nucletron International B.V.

Agreement States license these brachytherapy devices. Using these devices, a small, thin wire or sealed needle containing radioactive material, such as iridium-192 (Ir-192) or iodine-125 (I-125), is inserted directly into the cancerous tissue. The radiation from the isotope attacks the tumor as long as the device is in place. When the treatment is complete, long-lived material (Ir-192) is removed, but the short-lived radioisotopes (I-125) may be left permanently. This technique is used frequently to treat mouth, breast, lung, and uterine cancer.

Brachytherapy and teletherapy procedures are performed only in hospitals or clinics by trained medical personnel. Strict controls and safety requirements set by the NRC or the Agreement States must be followed. For example, treatment rooms must have adequate shielding to prevent scattered radiation from penetrating into an adjacent room. Radiation monitors must be used and patients carefully observed at all times during treatment.

Many types of cancer, such as Hodgkin's disease (cancer of the

The regulation and use of radioisotopes in today's world

lymph glands) and cancers of the cervix, larynx, and skin, can be treated by radiation alone. Boron capture neutron therapy has been used on a trial basis recently to treat potentially fatal brain cancer. In this procedure, the diseased brain tissue incorporates a neutron-absorbing isotope and then is exposed to neutron radiation originating from a nuclear research reactor. The energy and radiation emitted as a result of the neutron activation slow down the growth of cancer cells and, in some cases, completely kill them.

The overall objectives of NRC's safety rules for radiation medicine are to ensure that patients receive only the exposure medically prescribed and that the radiation is delivered in accordance with the physician's instructions.

NRC regulations require that physicians and physicists have special training and experience to practice radiation medicine. The training emphasizes safe operation of nuclear-related equipment and accurate recordkeeping.

When using radiation as a medical treatment, the physician weighs the potential benefits against the risk of side effects. Intense radiation exposure often destroys tumors that would prove fatal, but side effects such as hair loss,

The regulation and use of radioisotopes in today's world

How Are Radioisotopes Used in Industry?

Well-logging device

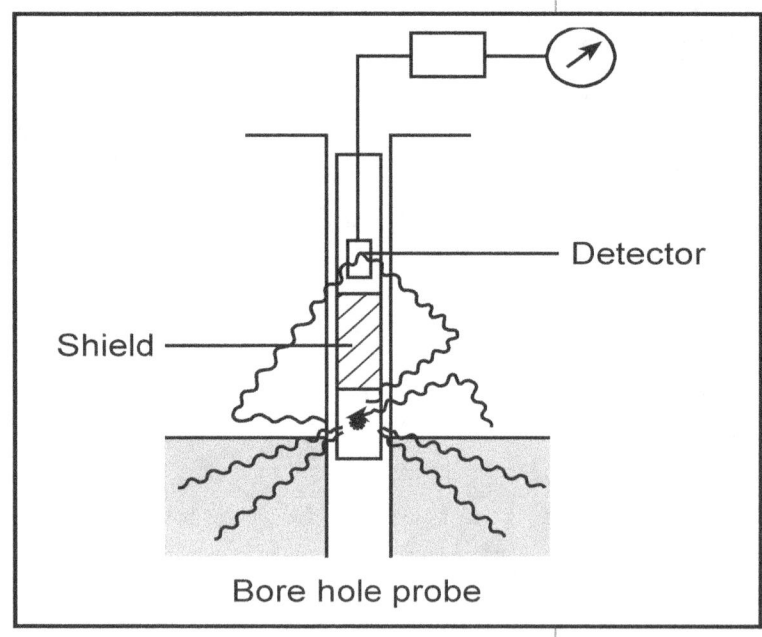

reduced white blood cell count, and nausea can be severe and must be monitored carefully.

Radioisotopes are used in many of today's industrial processes. High-tech methods that ensure the quality of manufactured products often rely on radiation generated by radioisotopes. To determine whether a well drilled deep into the ground has the potential for producing oil, geologists use **nuclear well-logging**, a technique that employs radiation from a radioisotope inside the well to detect the presence of different materials. Radioisotopes are also used to sterilize instruments; to find flaws in critical steel parts and welds that go into automobiles and modern buildings; to authenticate valuable works of art; and to solve crimes by spotting trace elements of poison, for example. Radioisotopes can also eliminate dust from film and compact discs as well as static electricity (which may create a fire hazard) from can labels.

The regulation and use of radioisotopes in today's world

Radiation can destroy germs, bacteria, and other harmful organisms that contaminate medical supplies, blood supplies, and even our food. Over the years, we have learned how to safely treat these items with radiation to sterilize or preserve them when other methods such as boiling or using chemical preservatives are either impractical or not as effective.

Medical supplies, such as rubber gloves, cloth bandages, syringes, and contact lens solution, can be sterilized by using an **irradiator.** An irradiator exposes the materials to gamma radiation, usually from Co-60.

Bacteria in foods can be destroyed by using radiation. Astronauts, for example, take food into space that often has been preserved using radiation. Food irradiation is discussed in greater detail later in this brochure.

The NRC regulates irradiators to protect workers and the public but, with some exceptions, does not specify the types of products that may be irradiated. The NRC does not evaluate the quality of irradiated products or the

Irradiation and Treatment of Materials With Radioisotopes

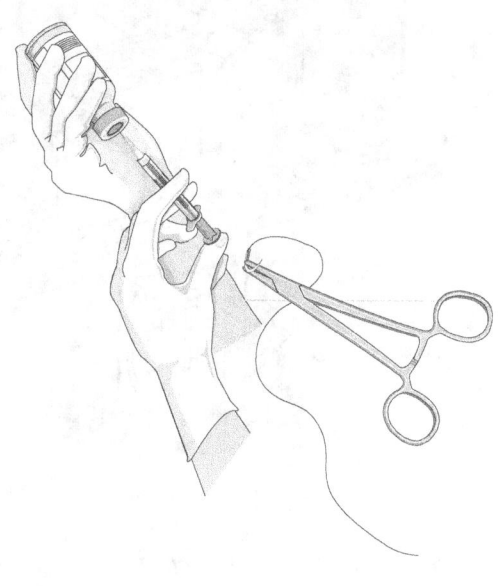

The regulation and use of radioisotopes in today's world

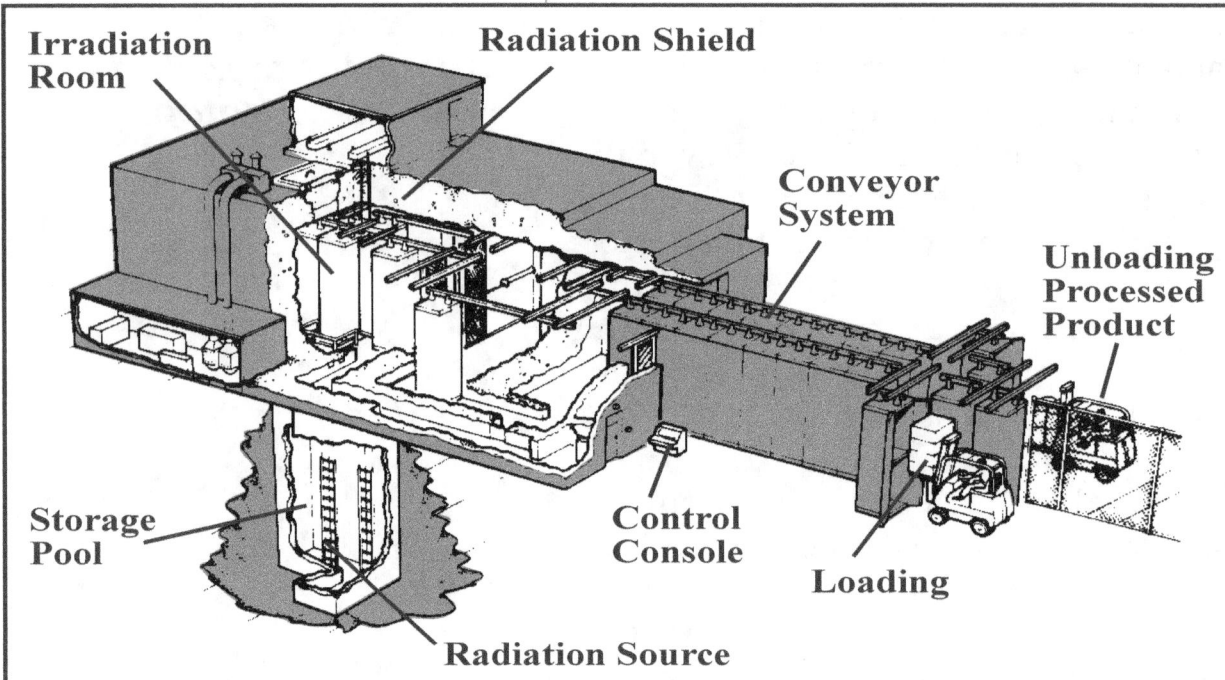

An industrial irradiator used for food products consists of a room with concrete walls 2 meters thick and a radiation source (Co-60). A conveyor system automatically moves the products into the room for irradiation and then removes them. When it is necessary for personnel to enter the room, the source is lowered to the bottom of the pool, where water absorbs the radiation energy and protects the workers.

Diagram: Courtesy of Food Technology Service, Inc.

technical or economic feasibility of product irradiation.

In the United States, food safety is regulated by the FDA and the U.S. Department of Agriculture (USDA).

There is an important difference between exposing materials to radiation

The regulation and use of radioisotopes in today's world

from an irradiator and exposing them to radiation from a reactor. The gamma radiation from Co-60 in an irradiator kills bacteria and germs, but does not leave any radioactive residue or cause any of the materials exposed to become radioactive themselves. This is because the Co-60 is contained in stainless steel capsules and does not commingle with the product being irradiated. This differs, however, from the effects of neutron radiation. Material **can** become radioactive after exposure to neutrons from a reactor or linear accelerator because the neutrons are absorbed by the nucleus of the atoms in the material, which then form different radioactive isotopes or elements (also referred to as **neutron activation**).

In manufacturing, radiation can change the characteristics of materials, often giving them features that are highly desirable. For example, wood and plastic composites treated with gamma radiation are used for some flooring in high-traffic areas of department stores, airports, hotels, and churches, because they resist abrasion and ensure low maintenance. Some gemstones, such as blue topaz, can be given a more pleasing color using neutron radiation treatment.

The regulation and use of radioisotopes in today's world

Gauge, luminous watch dial and smoke detector

Although there might be some residual radiation from neutron bombardment of gemstones and other consumer products, the half-life of these materials is generally so short (seconds or hours) that risk to the public is extremely low. *To ensure that the products are manufactured correctly, manufacturers and distributors of these materials must be licensed by the NRC.*

Today, luminous watch dials are painted with tritium or promethium-147 replacing radium used many years ago. These radioisotopes trigger a chemical reaction with other materials upon contact, which then give off a "glow." These materials are sometimes placed in gun sights or clocks and watches so they can be used at night. One frequently sees "exit" signs illuminated by tritium as safety markers on passenger aircraft, aboard ships, and in many buildings. Some gas lantern mantles are manufactured with the radioisotope thorium to provide a bright light when burned.

Measuring and Testing Applications

As radiation from a radioisotope passes through matter, it is scattered or absorbed. The amount of radiation passing completely through depends on the thickness and density of the matter. This property makes radiation useful for many manufacturing quality control processes.

The regulation and use of radioisotopes in today's world

For example, radioisotopes in gauges are used to monitor and control the thickness of sheet metal, textiles, aluminum foil, newspaper, copier paper, plastics, and photographic film as they are manufactured. Radiation penetrates the material as it is processed, and detectors measure the amount of radiation passing through and compare it to the amount that should be detected for the desired thickness of material. If the readings are too high or too low, controls in the manufacturing process can be activated to correct the problem. This is accomplished with a high degree of accuracy, and the materials need not be touched by human hands.

Radioactive tracers can also be used in the machine-tools industry to measure wear on cutting tools and drills. Gauges also play a role in the measurement of everyday objects, such as the amount of glue on a postage stamp, the

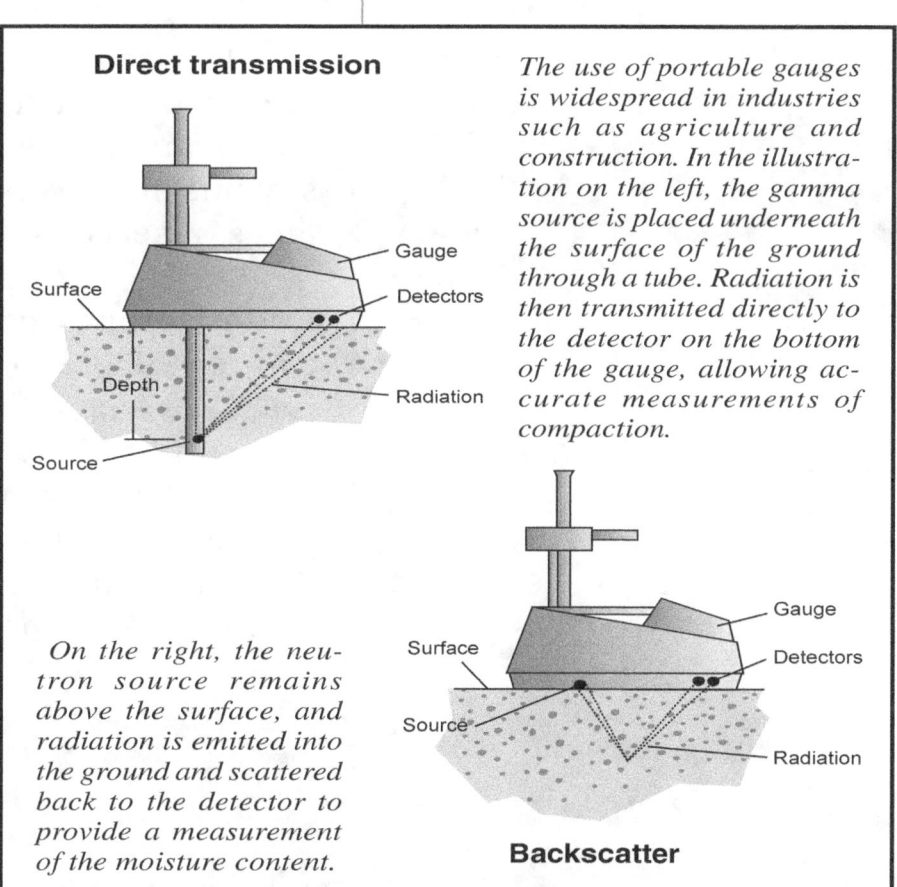

On the right, the neutron source remains above the surface, and radiation is emitted into the ground and scattered back to the detector to provide a measurement of the moisture content.

The use of portable gauges is widespread in industries such as agriculture and construction. In the illustration on the left, the gamma source is placed underneath the surface of the ground through a tube. Radiation is then transmitted directly to the detector on the bottom of the gauge, allowing accurate measurements of compaction.

The regulation and use of radioisotopes in today's world

Portable gauge used to test the structural integrity of roads and bridges

level of liquid in canned beverages, the amount of air whipped into ice cream, and the amount of tobacco packed into a cigarette. Other gauge and nuclear instrument applications include: (1) monitoring the structural integrity of roads, buildings, and bridges; (2) exploring for oil, gas, and minerals; and (3) detecting explosives in luggage at airports. To ensure adequate water supplies, radioisotopes are used to measure water runoffs from rain and snow, flow rates of streams and rivers, and leakage from lakes, reservoirs, and canals.

Radioisotopes can also be used to test for the presence of certain materials. For example, a small amount of americium-241 is found in most smoke detectors used in homes and offices. This radioisotope is part of the sensing unit that triggers an alarm when smoke is present.

The regulation and use of radioisotopes in today's world

Industrial Radiography

A process called **industrial radiography** is used to inspect many metal parts and welds for defects. A sealed radiation source, usually Ir-192 or Co-60, beams radiation at the object to be checked. Special photographic, or radiographic, film on the opposite side of the source is exposed when it is struck by radiation passing through — like an x-ray. More radiation will pass through if there are cracks, breaks, or other flaws in the metal parts and will be recorded on the film. Therefore, structural problems can be detected by studying the film.

Radiography is used throughout industry. For example, jet engine turbine blades are radiographed to ensure that internal cooling passages are manufactured properly.

The NRC or Agreement States license radiographers and periodically inspect them to ensure they follow safety procedures.

Portable radiography camera used to test for cracks in metal

How Are Radioisotopes Used in Agriculture?

Radioisotopes have played an important role in improving agricultural science. Radioisotopes are used as a research tool to develop new strains of food crops that are more resistant to disease, are of higher quality, allow earlier ripening, and produce a higher yield. When insects are sterilized with radiation, they mate without producing offspring, thus providing some control of insect population growth. With fewer pests, food crops can flourish. This technique has eliminated screw worm infestation in the southeastern United States and Mexico and has helped control the Mediterranean fruit fly in California. It has also been used to control mosquitoes and flies throughout the world.

Radioisotope tracers in plant nutrients enable scientists to make discoveries that have increased the effectiveness of fertilizer.

A major factor in successful crop production is the presence of an adequate water supply. Nuclear moisture density gauges can monitor the moisture content of soil, helping make the most efficient use of limited water sources.

The NRC and the Agreement States license byproduct radioactive materials used in these agricultural applications in the United States.

Radiation can be used to destroy bacteria in food and control insect and parasite infestation. Since 1963, the FDA has approved irradiation as a method to inhibit sprouting and to delay ripening in many fresh fruits and vegetables. The FDA has more recently approved the use of irradiation to control insects in food, trichina in pork, and bacteria in poultry, spices, and seasonings.

Currently, irradiation is used in the United States on very few food items; however, some spices are irradiated. In 1992, the USDA also approved the commercial irradiation of poultry, and in 1999, approved commercial irradiation of meat.

The FDA has established regulations controlling irradiation of food. Irradiated food must clearly

Food Irradiation and Its Safety

The regulation and use of radioisotopes in today's world

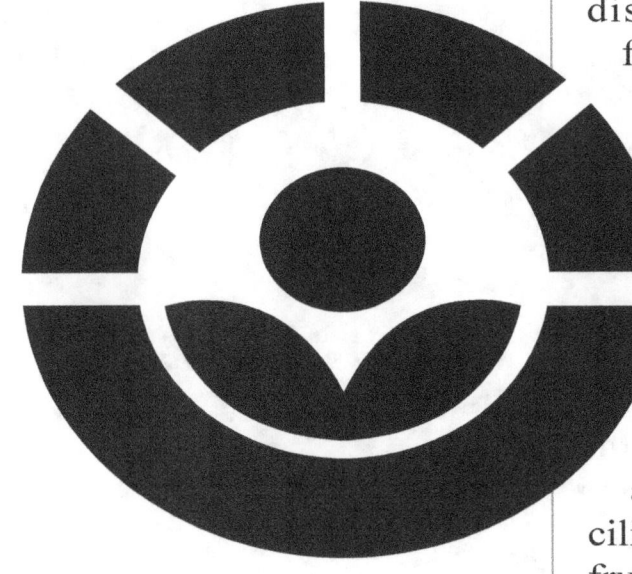

Radura logo

display the radura logo (a green, flower-like, international symbol for irradiation) and words such as "treated with radiation."

Governments in about 37 countries have approved irradiation of some 40 foods. In this country only one commercial facility is devoted solely to irradiating food products. It is located in Florida and licensed by the State. That facility irradiates strawberries and citrus fruits by exposing them to gamma rays from a radioactive source. When not in use, the radiation sources are safely stored in a pool of water.

Although it is a relatively new commercial process, food irradiation has been studied at length for over 40 years. According to the FDA, research to date has turned up no evidence of adverse health effects from irradiation of the treated food. The radiation passes through the food leaving no radioactive residue. Some concerns have been raised regarding the chemical changes that may alter the finished food product. However, research has shown that changes in flavor and texture caused by irradiating food are similar to changes due to canning or freezing. Studies are continuing to ensure that all irradiated food is safe to eat.

The regulation and use of radioisotopes in today's world

What Are Other Uses of Radioisotopes?

Geology, Archeology, and Space

Through a method called **radiocarbon dating**, archaeologists can determine when formerly living materials were last alive. This procedure relies on the presence of C-14, a naturally occurring, long-lived radioisotope present in all living things.

The ratio of C-14 to C-12 in the atmosphere has been relatively constant throughout history. When an animal or plant dies, it stops taking in carbon, and the amount of C-14 in its tissue begins to decrease through the process of radioactive decay. Comparing the C-14 to C-12 ratio in dead material with the "living ratio" enables us to calculate how long ago the plant or animal lived. This method was used to determine the approximate age of two major archaeological discoveries in recent decades: the Shroud of Turin and the Dead Sea Scrolls. C-14 analysis has also been used during space missions to test for life on Mars.

Heat generated by radioisotopes is used to power small generators used for remote applications, such as in space exploration.

Radioisotope-powered electrical generators have been used to power exploration space craft, navigational and weather satellites, and communication

The regulation and use of radioisotopes in today's world

THE ABCs OF RTGs

How does a radioisotope thermoelectric generator (RTG) produce electricity? The "fuel"—a plutonium isotope that can't explode—"decays," or loses its radioactivity, giving off heat. That heat is absorbed by a device called a **thermocouple**, which consists of a compound of two metals—silicon and germanium—that are "doped." "Doping" the metals—adding impurities to them—changes their electrical and thermal properties. As a result, energy moves from the "hot side" of the thermocouple— the side closer to the fuel—to the "cold side," producing a voltage. Gas circulating through cooling tubes keeps the RTG from getting too hot, and fins dissipate the heat.

The thermocouple was first demonstrated early in the last century, but the efficiency was too low to be practical. In the 1950s, researchers discovered they could get higher efficiencies from a thermocouple by using "semiconductor" materials such as silicon and germanium. After that, there was no looking back. Today's conversion efficiency is about 5 percent, but with some of the conversion technologies now under development, tomorrow's efficiencies are expected to be at least 20 percent.

Height: 45 inches   Diameter: 18 inches   Weight: 123.5 pounds

Diagram: Courtesy of the Nuclear Energy Institute

satellites. They allow us to operate weather stations at the North and South Poles, seismic sensing stations in remote locations, and devices placed on the ocean floor for scientific investigations and national defense.

The Department of Energy (DOE) manufactures radioactive thermal generators, which are powered by plutonium-238 and Sr-90. The NRC has provided technical assistance for some of these applications.

A procedure called **activation analysis** is frequently used to identify trace quantities of materials such as glass, tape, gunpowder, lead, and poisons that are important to criminal investigations. Samples of

The regulation and use of radioisotopes in today's world

materials are placed in a nuclear reactor and bombarded with neutrons. The induced radiation produces a "fingerprint" of the elements in the sample.

In addition, activation analysis can prove the authenticity of old paintings by detecting whether certain modern materials are present. Museums routinely use this and other techniques that rely on radioisotopes to spot forgeries.

Radioisotopes are used to detect pollution in the air, water, and soil. Sources of acid rain and greenhouse gases that may be causing global warming can be traced using radioisotopes. In addition, electron beam radiation can be used to remove gaseous pollutants such as sulphur dioxide or nitrogen oxide, both harmful gases emitted from coal-fired power plants and other industries.

When used properly, radioisotopes are a productive part of today's world. The NRC and the Agreement States remain committed to protecting public health and safety in the use of these nuclear materials by inspecting medical, academic, and industrial applications carefully, and monitoring users to ensure safe practices.

Conclusion

The regulation and use of radioisotopes in today's world

Appendix: Some of the Major Uses of Radioisotopes in the United States.

Americum-241

Used in many smoke detectors for homes and businesses...to measure levels of toxic lead in dried paint samples...to ensure uniform thickness in rolling processes like steel and paper production...and to help determine where oil wells should be drilled.

Cadmium-109

Used to analyze metal alloys for checking stock, scrap sorting.

Calcium-47

Important aid to biomedical researchers studying the cellular functions and bone formation in mammals

Californium-252

Used to inspect airline luggage for hidden expolsives...to gauge the moisture content of soil in the road construction and building industries...and to measure the moisture of materials stored in soils.

Carbon-14

Major research tool. Helps in research to ensure that potential new drugs are metabolized without forming harmful by-products. Used in biological research, agriculture, pollution control, and archeology.

Cesium-137

Used to treat cancerous tumors...to measure correct patient dosages of radioactive pharmaceuticals...to measure and control the liquid flow in oil pipelines...to tell researchers whether oil wells are plugged by sand...and to ensure the right fill level for packages of food, drugs and other products. (The products in these packages do not become radioactive.)

Chromium-51

Used in research in red blood cell survival studies

Cobalt-57

Used as a tracer to diagnose pernicious anemia

Cobalt-60

Used to sterilize surgical instruments...and to improve the safety and reliability of industrial fuel oil burners. Used in cancer treatment, food irradiation, gauges, and radiography.

Copper-67

When injected with monoclonal antibodies into a cancer patient, helps the antibodies bind to and destroy the tumor.

The regulation and use of radioisotopes in today's world

Curium-244

Used in mining to analyze material excavated from pits...and slurries from drilling operations.

Gallium-67

Used in medical diagnosis

Iodine-123

Widely used to diagnose thyroid disorders and other metabolic disorders including brain function.

Iodine-125

Major diagnostic tool used in clinical tests and to diagnose thyroid disorders. Also used in biomedical research.

Iodine-129

Used to check some radioactivity counters in in vitro diagnostic testing laboratories.

Iodine-131

Used to treat thyroid disorders. (Former President George Bush and Mrs. Bush were both successfully treated for Graves' disease, a thyroid disease, with iodine-131.)

Iridium-192

Used to test the integrity of pipeline welds, boilers and aircraft parts and in brachytherapy/tumor irradiation.

Iron-55

Used to analyze electroplating solutions and to detect the presence of sulphur in the air. Used in metabolism research.

Krypton-85

Used in indicator lights in appliances such as clothes washers and dryers, stereos, and coffeemakers...to gauge the thickness of thin plastics and sheet metal, rubber, textiles and paper... and to measure dust and pollutant levels.

Nickel-63

Used to detect explosives, and in voltage regulators and current surge protectors in electronic devices, and in electron capture detectors for gas chromatographs.

Phosphorus-32

Used in molecular biology and genetics research.

The regulation and use of radioisotopes in today's world

Phosphorus-33
Used in molecular biology and genetics research.

Plutonium-238
Has powered more than 20 NASA spacecraft since 1972.

Polonium-210
Reduces the static charge in production of photographic film and other materials

Promethium-147
Used in electric blanket thermostats...and to gauge the thickness of thin plastics, thin sheet metal, rubber, textile and paper.

Radium-226
Makes lighting rods more effective.

Selenium-75
Used in protein studies in life science research.

Sodium-24
Used to locate leaks in industrial pipe lines...and in oil well studies.

Strontium-85
Used to study bone formation and metabolism.

Strontium-90
Used in survey meters by schools, the military and emergency management authorities. Also used in cigarette manufacturing sensors and medical treatment.

Sulphur-35
Used in genetics and molecular biology research.

Technetium-99m
The most widely used radioactive pharmaceutical for diagnostic studies in nuclear medicine. Different chemical forms are used for brain, bone, liver, spleen and kidney imaging and also for blood flow studies.

The regulation and use of radioisotopes in today's world

Thallium-201

Used in nuclear medicine for nuclear cardiology and tumor detection.

Thallium-204

Measures the dust and pollutant levels on filter paper...and gauges the thickness of plastics, sheet metal, rubber, textiles and paper.

Thoriated Tungsten

Used in electric arc welding rods in construction, aircraft, petrochemical and food processing equipment industries. They produce easier starting, greater arc stability and less metal contamination.

Thorium-229

Helps fluorescent lights last longer.

Thorium-230

Provides coloring and fluorescence in colored glazes and glassware.

Tritium

Major tool for biomedical research. Used for life science and drug metabolism studies to ensure the safety of potential new drugs...for self-luminous aircraft and commercial exit signs...for luminous dials, gauges and wrist watches...to produce luminous paint, and for geological prospecting and hydrology.

Uranium-234

Used in dental fixtures like crowns and dentures to provide a natural color and brightness.

Uranium-235

Fuel for nuclear power plants and naval nuclear propulsion systems ...and used to produce fluorescent glassware, a variety of colored glazes and wall tiles.

Xenon-133

Used in nuclear medicine for lung ventilation and blood flow studies.

Courtesy of Management Information Systems, Inc.

Notes

U.S. Nuclear Regulatory Commission

Washington, DC 20555-0001
Office of Public Affairs

NUREG/BR-0217 Rev. 1
APRIL 2000

www.ingramcontent.com/pod-product-compliance
Lightning Source LLC
Chambersburg PA
CBHW081807170526
45167CB00008B/3365